STOTHERT & PITT

Cranemakers to the World

THE
Stothert & Pitt Magazine

STOTHERT & PITT

Cranemakers to the World

KEN ANDREWS AND STUART BURROUGHS

TEMPUS

Frontispiece: DD2 Cranes at the Port of London, 1964.

First published 2003

Tempus Publishing Limited
The Mill, Brimscombe Port,
Stroud, Gloucestershire, GL5 2QG

British Library Cataloguing in Publication Data.
A catalogue record for this book is available from the British Library.

ISBN 0 7524 2794 6

Typesetting and origination by Tempus Publishing Limited
Printed in Great Britain by Midway Colour Print, Wiltshire

Contents

Introduction

In the summer of 1988, shortly before the engineering firm of Stothert & Pitt closed its manufacturing plant at Bath, the company's collection of photographic negatives and prints were deposited at the Museum of Bath at Work.

The majority of the photographs in this book have been chosen from this unique archive. This collection of images was taken by the company's own full-time photographic department and chronicle the manufacture, assembly and installation of practically every product from the late nineteenth century onwards. In addition, photographs for the staff magazine, *Stothert & Pitt Magazine*, were stored and these show staff, workplaces and activities, often in candid detail. Owing to the loss of much of the company's paperwork archives through bomb damage in the Second World War and the persistent river flooding of the factory sites, this collection represents some of the only evidence in Bath of the operation's existence. These illustrations provide vital historical detail to accompany that information which is known about the company's activities from printed catalogues and interviews with former employees. Stothert & Pitt's factory sites in Bath have been redeveloped or completely demolished and only a handful of their products may be seen in the city. The collection of photographs is largely on glass plates, some up to 10in x 12in in size and, as a result of inappropriate storage at Stothert & Pitt, there has been some loss of definition of some of the prints.

When Stothert & Pitt Co. Ltd announced the closure of its engineering works in January 1989, 580 local people lost their jobs and the city of Bath lost its single largest manufacturer. For over 200 years the heavy engineering and metalworking business had employed local people (over 2,000 in 1945) and supplied a wide variety of products from bedsteads to boilers and cement mixers to cranes.

Today, when Bath is renowned for an 'industry-free' elegance, refined and displayed for tourists, this opportunity to remind that heavy industry has played its part in the city's evolution could not be more timely. In 1980 the director of the Science Museum claimed that the contribution by Stothert & Pitt as a supplier of heavy engineering across the world was Bath's greatest contribution to world history. This apparently surprising remark reveals an acknowledgment of the company's significance, locally, nationally and internationally. Many former workers at the firm have assisted in the production of the text for this book and it is to them and the thousands of workers employed by the business over 204 years that this book is dedicated.

Acknowledgements

We would like to express our appreciation for the help by former employees of Stothert & Pitt in the production of this book. Particularly Ian Guy and Mike Stockley. We are also grateful to the staff of the Bath & North East Somerset Archive Office, Peter English, Don Browning, Valerie Andrews and especially Dr Hugh Torrens, without whose work this book would not have been possible.

References

Harper, Duncan, *Bath at Work* (Millstream Books, 1989)
Torrens, Hugh S., *The Evolution of a Family Firm – Stothert and Pitt of Bath* (Stothert & Pitt Ltd, 1978)
Torrens, Hugh S., 'Ernest Feuerheerd and his Rotary Lobe Pump of 1919' (MS)
Mike Chapman and Elizabeth Holland, Journal 30 of the Bristol Industrial Archaeological Society, pp.34–38 (1977)
The collection of Stothert & Pitt magazines, photographs and miscellanea at the Museum of Bath at Work
The Stothert & Pitt Collection at the Bath & North East Somerset Archive Office

The Iron Foundry, Torrance & Sons Works, Bitton, c.1925. Over 40,000 images were recorded by Stothert & Pitt's photographic department.

15-ton block-setting crane from an 1885 Stothert & Pitt catalogue:

The machine which we illustrate has been constructed for Port Alfred Harbour, this being one of several harbours now being made by Sir J. Coode in South Africa. The pier, for the construction of which the crane will be employed, will consist of concrete blocks laid on what is known as the 'over-end system'. The machine has been specially designed throughout, and represents the most complete development which block setting plant has yet attained.

    The most striking features of the crane are, the great range of all the motions, the large radius and the method of providing for the latter by a horizontal jib suspended from a king post. The principal dimensions of the crane are, total height of lift 46ft, radius variable from 15ft minimum to 45ft maximum, height from rail to underside of jib 22ft 2¾in, radius of tail to centre of boiler 312ft, working load 15 tons, proof load 19 tons.

one

History

To find the historical roots of the engineering giant that Stothert & Pitt became in the elegance of late eighteenth century Bath might at first seem surprising. Certainly the 'economic miracle' of the eighteenth century, which had transformed the city from a sleepy provincial town to a cosmopolitan city of fashion, had largely been in the service economy, supplying visitors and new residents with a high standard of accommodation. However, the almost continuous growth of the city, with periodic building booms during the same period, had also created a great demand for all manner of construction materials and metalwork (and metalworkers) of all kinds. It is against the backdrop of feverish building activity that ironmongers such as George Stothert flourished.

George Stothert had arrived in Bath from Shropshire around 1774 to work at Thomas Harris's ironmongers business in Horse Street (part of what is now Southgate Street). The firm also employed 'smiths, braziers, tinmen, planemakers, etc.' In 1785 Stothert took control of the operation as one of the most vigorous building booms was beginning in the city.

In 1795 he was registered as 'George Stothert, Ironmonger, Smith, Brazier, Tin Man, Plane Maker, also operator of a Manufactures Register and Supplies of all kinds of stove grates', and in 1796 Stothert cemented his success by becoming a major outlet for products by Abraham Darby's Coalbrookdale Co. His family contacts in Shropshire may have assisted in this business connection, but it is testament to George Stothert's business acumen that he saw the advantage of such a relationship with such a revolutionary concern. Within a few years the business was supplying domestic ironmongery of all kinds, including fire grates, ovens, gates, staircases, cast-iron pipes, banisters, frying pans and iron bookcases. In 1799 Stothert supplied two cast-iron bridges to span the newly constructed Kennet & Avon Canal as it cut across Sydney Gardens (or Vauxhall and Ranelagh Gardens as they were called) which were erected a year later at the total cost of £456 1s 6d. The canal company was a major customer and between 1799 and 1807 nearly £1,000 worth of castings was supplied from the Coalbrookdale Co.

By 1798 Stothert & Co. owned a warehouse on separate premises in Northgate Street where in 1800, during food riots in the city, 'an atrocious attempt [was] made to fire the premises of Mr Stothert, one of the most respectable and publick [sic] spirited and esteemed inhabitants of Bath'. In the following year Stothert invested in the woollen mills at Twerton and Sydney Gardens; he also acted as agent for a portable brewing machine invented by James Needham and established an international reputation for fine woodworking planes.

In 1815, prompted by both the Coalbrookdale Co.'s decision to open its own warehouse in nearby Bristol and the ending of the Napoleonic Wars, George Stothert's son, also named George, set up an iron foundry alongside the ironmongery business in Horse Street. A range of goods similar to those supplied from Shropshire were offered, including bread ovens and fire grates.

By 1827 Messrs Stothert were described as 'Furnishing Ironmongers to His Majesty' and the business continued to expand with the iron foundry and engineering works (which by this time had expanded into nearby Philip Street) renamed as the Newark Foundry, equipped with steam power and supplying drain gratings, agricultural equipment and water pumps.

Evidence of the company's success is shown when, in 1836, the family set up a separate engineering company in Bristol for the manufacture of steam locomotives and, eventually, steamships.

The demands of the railway construction industry and the continuous expansion of maritime trade had created opportunities for engineers in cargo handling, heavy lifting and civil engineering equipment and, during the next 100 years, Stothert & Pitt would exploit this demand. At the Great Exhibition of 1851 the firm exhibited an early hand crane and, through a multitude of products supplied during the rest of the century (including gas apparatus, pumps, heating, cooking and washing apparatus), crane manufacture began to dominate.

General engineering work continued, however, and in 1855 the company supplied water-pumping equipment for prefabricated hospitals designed by I.K. Brunel for the Crimean War. In 1857 the company began moving to larger and purpose-built premises, the Newark Works, nearby on the Lower Bristol Road.

From the 1850s steam-powered cranes were supplied and in 1876 a 35-ton lifting crane, to a design improved by engineer William Fairbairn, was supplied to Bristol City Docks. In 1890 a second large factory, the Victoria Works, was laid out.

A catalogue of 1885 gives an idea of the range of manufactures on offer, with 1,050 separate items advertised. By this time these included a range of lifting gear, including giant cranes. The development of concrete had stimulated harbour construction, and one range of cranes in particular, the Titan class, was designed specifically for laying concrete blocks up to 50 tons in weight, as part of artificial harbours. Deck cranes and machinery were also supplied, including winches and capstans.

Electrically operated dockside cranes were first supplied to Southampton Harbour Commissioners in 1892 and deck equipment was supplied to Harland & Wolff for installation on the *Titanic* in 1911.

The expansion of the company continued and in 1898 Stothert & Pitt Co. Ltd became a public registered company. Inventions at this time included development of Claude Toplis's 'level luffing' mechanism in 1912, which gave the company's cranes an instant advantage over their competitors.

The First World War created a demand for military hardware; a primitive tank was developed by 1916, along with deck and harbour equipment for the Royal Navy. In the interwar years the firm of Torrance & Son, engineers at nearby Bitton, was acquired (1919) and the manufacture of the Fuerheerd rotary pump (1920) and positive displacement pumps (1936) began. The business flourished, supplying foreign governments and the Crown Agents for Colonies with cargo-handling equipment.

The Second World War created an even greater demand for a range of lifting and pumping equipment and in 1942, with direction from the Admiralty which had moved to Bath for the duration of the war, the company developed a top-secret mini submarine.

After the war expansion continued at the Victoria Works, and in the 1960s a new apprentice training school and depots in Birmingham and Isleworth were opened. Former textile mill buildings were acquired in Twerton and a new office block in Oak Street opened in 1968. In 1969 the Bridgwater crane-making firm of W&F Wills was acquired.

By the 1970s the workforce numbered around 1,800 people and, having won design awards for its cranes in 1968 and 1970, the company's future seemed assured. However, a combination of the worldwide economic depression of the 1970s and increasing competition from abroad forced the company to contract.

Despite winning the Queen's Award for Export Achievement in 1979, the firm was sold to financier Robert Maxwell's Hollis Group in 1986. In 1988 a management purchase from Hollis failed to prevent the firm closing in January 1989. The name Stothert & Pitt continues as a design and consultancy operation run through Langley Holdings PLC.

Invoice to George Stothert, 1795.

Canal Bridge in Sydney Gardens, Bath. George Stothert supplied two bridges from the Coalbrookdale Co. to the Kennet & Avon Canal Co., which were erected in 1800.

NOTICE is hereby given, that the PARTNER-SHIP lately subsisting between us, GEORGE STOTHERT the elder and GEORGE STOTHERT the younger, of the city of Bath, Furnishing Ironmongers and Manufacturers, under the firm of Stothert and Son, was on the 24th day of June last DISSOLVED by mutual consent; and that all debts due and owing from us in respect of our late concern, are to be paid by the said George Stothert the elder, to whom all persons indebted to the said concern, are to pay the amount of their respective debts: And that the same trade or business hath since and will continue to be carried on by the said George Stothert the elder, on his own separate account. As witness our hands this 31st day of August, 1815.

GEORGE STOTHERT, Sen.

1473]   GEORGE STOTHERT, Jun.

GEORGE STOTHERT, Jun. respectfully informs his Friends and the Public, that he has established at No. 17, HORSE-STREET, in this City,

An IRON-FOUNDRY;

Where he casts and fits up the following articles, on the most reasonable terms:

Furnaces, Furnace Doors and Bars, Cisterns, Troughs, Pipes of all sizes, Columns, Street Posts, Pallisading, Field and Garden Rollers, Bookcases and Chests, Ovens, Oven Doors and Frames, Forge Backs, Plates, Stove Metal, Gratings, Scale Weights, Sash Weights, Wheels, and Machinery, to patterns of every description.

Bath, Sept. 1815.   [1474

Advert in *Bath & Cheltenham Gazette*, 1815. This advert announces the withdrawal of George Stothert Senior from the business and the transfer of control to his son, George. In addition, the opening of the new iron foundry is reported.

Stothert bread oven door. A large number of bread ovens and grates were manufactured and supplied to Bath homes after the opening of George Stothert's iron foundry in 1815.

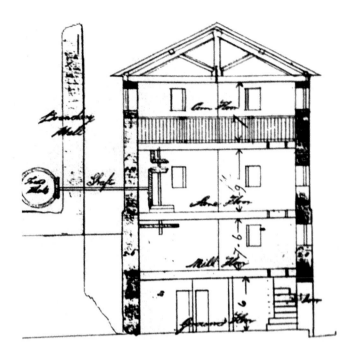

Treadmill and millhouse at Shepton Mallet, 1823. The Shepton Mallet 'House of Correction' contracted Henry Stothert to construct and equip a millhouse for grinding corn powered by a treadmill. The millhouse was outside the prison wall. The treadmill, powered by prisoners, was within the wall and drove the machinery via a shaft which pierced the 'boundary wall'. The treadmill was one of the largest installed in a British prison.

G. K. STOTHERT & CO.,

ENGINEERS, BOILER MAKERS AND IRON SHIP BUILDERS,

BRASS AND IRON FOUNDERS, COPPERSMITHS, &c.,

BRISTOL.

Right: Fire grate by John and William Stothert, *c.*1820. Between 1819 and 1841 John and William Stothert operated as Messrs Stothert and manufactured household ironmongery, grates, ranges and other goods.

Below: Stothert Bridge over the Kennet & Avon Canal at Widcombe, Bath. Stothert's iron foundry supplied this small footbridge for the highest lock on the Widcombe 'flight' of locks on the Kennet & Avon Canal. In the 1970s Stothert & Pitt restored the bridge. It is probably the oldest substantial product of the company. A second footbridge was also supplied for the Wash House lock.

Opposite below: The Shipyard of G.K. Stothert & Co. Ltd, Hotwells, Bristol, 1897. By 1837 Henry Stothert had established a separate Bristol operation undertaking locomotive and marine steam engine production to their own designs. By 1852 they had moved to this site and they continued until 1933 as a separate business.

Left: Mr Robert Pitt (1818-1886). Pitt joined Henry Stothert's company as an apprentice around 1834 and became a partner in the company, along with George Rayno, in November 1844.

Below: Box Tunnel for the Great Western Railway, Wiltshire. The huge task of constructing the 2,418ft-long railway tunnel, through the Wiltshire hills, had been contracted by I.K. Brunel to two local building firms, who were used to working through Bath limestone. Henry Stothert almost certainly supplied water pumps for the project, one reported to have pumped 1,200 gallons a minute. Lifting equipment may also have been supplied. The line through the tunnel was complete in 1841.

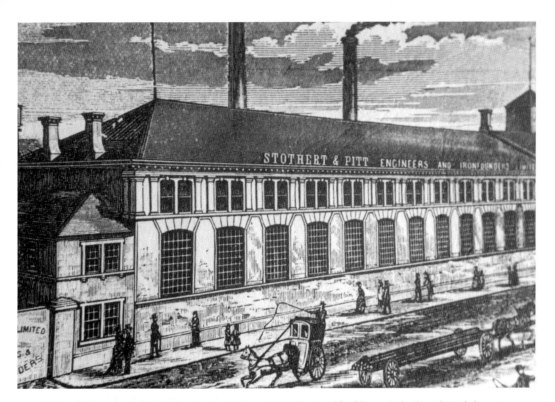

Newark Foundry, c.1860. The façade of this architect-designed building was built with Bath limestone with red brick behind. This view shows the extent of the machine shops as originally built. The artist has exaggerated the width of the adjacent Lower Bristol Road. The sheet metal sign on the parapet was designed to be visible from the Great Western Railway line on the embankment opposite.

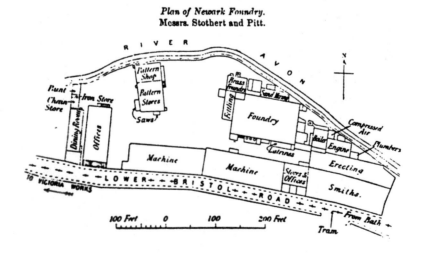

Plan of Newark Foundry, 1908. The opening of the new Newark Foundry in 1857 provided greatly enhanced facilities for the company. This plan accompanied a report on a visit to the works by the Institute of Mechanical Engineers in 1908.

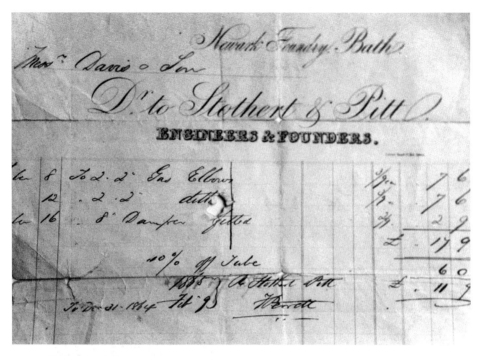

Invoice from Stothert & Pitt to Davis & Sons, 1864. Davis & Sons was a small local firm of ironworkers and were often contracted by Stothert & Pitt to undertake work for them.

Floods in Lower Bristol Road, 1925. Throughout its history the flooding of the River Avon had caused difficulties for Stothert & Pitt; not only had it held up business but it destroyed paper records. Frequent and serious flooding was finally prevented by improvements to the river alignments in 1968.

John Lum Stothert (1829-1891). John Lum Stothert was the first works manager (with Robert Pitt) of the new Newark Foundry in 1857. He continued to work as chairman of the company until his death in 1891.

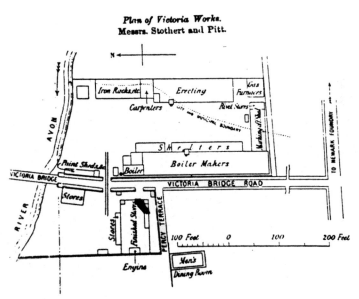

Plan of Victoria Works, 1908. In 1890 the Victoria Works was laid out with plenty of flat space for large-scale erection and fabrication work. The premises grew progressively during the twentieth century.

Aerial photograph of Victoria Works, 1975. The full extent of this large site can be seen in this photograph, with the long erecting sheds running parallel and at right angles to the River Avon. To the left of centre is the Victoria suspension bridge and originally the works were confined to the space above the road that ran to it. Expansion 'below' the bridge began in the 1920s.

Stothert & Pitt Rugby Football Club, Corston, Bath, 2003. Although Stothert & Pitt manufacturing ceased in 1989, its rugby club celebrated its 100th anniversary in 2003. The playing grounds are at Corston. On the sign can be seen the club emblem which appropriately gives the club its nickname – 'The Cranes'.

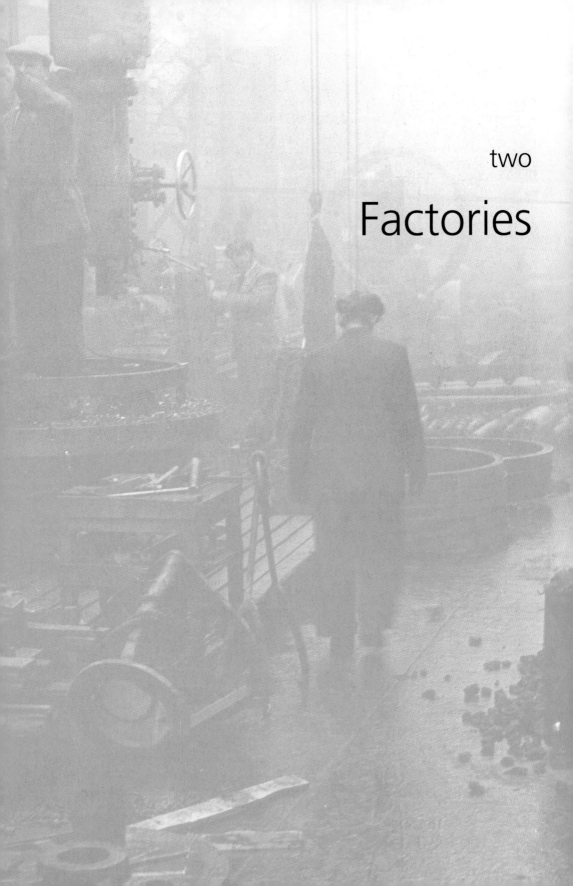

two

Factories

In 1785, when George Stothert took over the running of Harris's ironmongery concern, the business operated from a retail and storage premises in Horse Street. The expansion of business during the building boom in Bath of 1785-1793 and the Coalbrookdale Co. agency prompted Stothert to rent or lease a warehouse near George Stothert's home at 15 Northgate Street.

In 1815 the Horse Street premises were extended to accommodate the new iron foundry. By 1826 ironmongery, retail and storage was concentrated at Northgate Street, while the Horse Street foundry had expanded through to the road running to the east, Philip Street. By 1830 the Horse Street works had been renamed the Newark Foundry and the expansion of business and the size of products was putting strain on this site.

In 1857 the Newark Foundry moved to new purpose-built premises on the Lower Bristol Road and the Horse Street premises were sold. Designed by Bath-born architect Thomas Fuller, the new factory building (which still stands, although altered over the years) measured 300ft in length, with fitting and erecting shops 162ft long.

At the rear of the works a large open yard backed onto the river wharf with a travelling crane for erection in this yard.

The ironmongery business having been sold off in the 1840s, manufacturing continued to expand and in 1890 another large area of riverside land near the Midland Railway station and goods yard was laid out with new workshops.

The Victoria Works, as this factory was known, occupied the site of a former timber yard and was equipped with boiler-making workshops, structural steel department and large erecting shops, whilst the Newark Foundry housed the secretarial and office blocks, machine shops, pattern shop, foundry, smithy, tool room and a smaller erecting shop. The Victoria Works complex expanded during the early twentieth century, building over the former Percy Terrace, Longmead and Cross Streets, and in the 1950s Albert Terrace and a former cricket ground were absorbed. From 1898 until 1970 the works were connected by rail to the Midland Railway goods depot, with coal and raw material coming in and products taken out.

In 1954 former textile mill buildings at Weston Island were occupied for storage and testing of Vibroll Rollers. Other development in the Twerton area in the 1960s included the Avalon Apprentice Training School and workshops and storage premises.

All of the riverside sites were prone to regular flooding which caused havoc, often resulting in the complete closure of the operations for days on end.

In 1989 all the premises were closed and the large erecting sheds quickly dismantled at the Victoria Works. The buildings at the Newark Foundry were redeveloped for small business occupancy and the other buildings sold.

Aerial photograph of Newark Foundry, 1975. The cramped works site in the middle of this photograph is shown bounded on one side by the Lower Bristol Road and by the River Avon on the other. The large goods shed in the Great Western Railway yard used by Stothert & Pitt for transporting may be seen on the right.

Machine Shop at Newark Foundry, 1947. The original machine shop was planned in 1857 to include a high-level gallery, which by the time of this photograph had been converted for electrical works.

Newark Foundry Yard, 1925. This view of the flooded yard at the Newark site shows the machine shops on the right, the erecting shop in the distance and the foundry buildings on the left. The fettling shop stands before the foundry and a caterpillar crane used for shifting materials in the yard is stranded.

Above: Iron Foundry at Victoria Works, 1947. The traditional sand moulding of iron, brass and alloy castings was replaced with a mechanised installation at this site in the 1950s. Moulding boxes can be seen stacked on the floor.

Right: Entrance to Newark Foundry, 1947. The foundry was particularly susceptible to river flooding, rendering the works unusable for days on end. In this view the works yard can be seen beyond the entrance gates on the Lower Bristol Road. The inclined walkway links the administration offices on the left to the works on the right.

Left: Crawler crane used for erecting work at Newark Foundry Yard, 1933. In the background can be seen the office block.

Below: Machine Shop at Newark Foundry, 1960. The main machine shop, with the heaviest machinery, was situated at the Newark Foundry. On the right of this view stands a heavy-duty vertical milling machine.

Right: Newark Foundry Yard, 1924. In more clement weather, this view of the yard shows the machine shops behind the two light-loading cranes and the foundry on the right.

Below: Aerial photograph of Victoria Works, 1934. New boiler workshops were built at the Victoria Works in the 1920s and an area of terraced housing was purchased, cleared and developed. One row of houses, Albert Terrace, seen here on Victoria Bridge Road, survived until after the Second World War.

Right: A 50-ton Titan crane at Victoria Works, 1925. One reason for expanding onto the new Victoria Works site was to undertake the building of much larger cranes designed by the company from the 1880s. In the foreground of this view can be seen Victoria Bridge Road which ran through the works and the securing ties of the Victoria suspension bridge over the River Avon to the left.

New Boiler Shop at Victoria Works, 1925. The Victoria Works was rail-connected from the very start to the Midland Railway goods yard and a self-propelling rail-mounted crane of Stothert & Pitt's own manufacture is shown at the boiler shop entrance, shortly after this huge shed was completed.

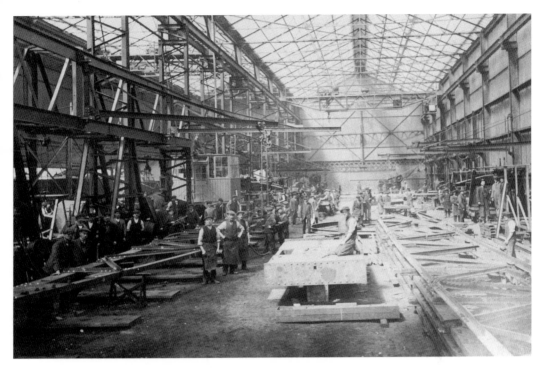

New Boiler Shop, Victoria Works, 1925. In the 1920s expansion continued at the Victoria Works with this new installation for fabricating boilers and structural components. A Stothert overhead travelling crane spans the 56ft width of the building and was capable of lifting loads up to 15 tons.

No.2 Machine Shop at the Victoria Works, 1952. The company's machine shops operated at the Newark Foundry, but such was the demand for machining, especially during the two world wars, that machine shops were opened at the Victoria Works as well.

Right above: Works canteen, 1960. Although canteens had been provided at the two main sites, a new facility opened at the junction of Lower Bristol Road and Victoria Bridge Road in the late 1950s. Once again Stothert facilities are seen closed due to flooding!

Right: Automatic vending machine, 1967. The year 1967 saw the introduction of vending machines into the offices at Stothert & Pitt. As might have been expected, of course, there were supporters and opponents to the new scheme.

New distribution depot at Birmingham, 1960. In the 1960s new sales and distribution depots were opened at Isleworth and at Birmingham. A London office in Victoria Street had operated since the nineteenth century.

Stothert & Pitt Avalon Apprentice Training School, 1980. In the 1960s an apprentice training facility was opened at Twerton, to the west of Bath. These truck-mounted cement mixers have left the storage facilities on Weston Island via a river bridge adjacent to the training school.

Opposite below: Oak Street offices, 1969. In May 1967 office staff moved from the Newark Foundry across the road into this new accommodation The office block was built on stilts to give the maximum amount of space for car parking and to avoid the inconvenience of flooding.

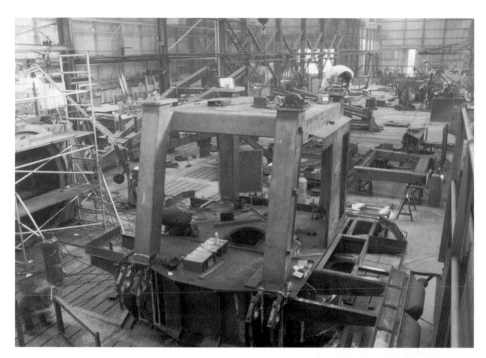

Fabrication workshop at Victoria Works, 1980. The erecting shops at the Victoria Works were used for construction and assembly of some of the largest company products, including the Titan and Goliath class cranes. In this recent view, part of a crane superstructure is under construction.

Spares store at Weston Island, Twerton, 1980. Former mill buildings associated with the nearby woollen industry had been acquired by Stothert & Pitt in the 1960s, although buildings had been in use there since the Second World War. In 1969 a purpose-built store was erected to supply the whole business.

Left: The new Contractors Plant Paint Shop at Weston Island in 1961.

Opposite below: This aerial view of Weston Island shows the concrete road completed in 1969. Materials are shown stored outside the newly-built storage and spares store. It was rumoured that Vibroll compactors were only tested on the island, as demonstrating them on dry land was likely to cause too much vibration!

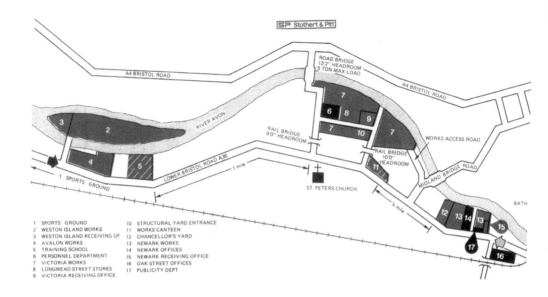

SP Stothert & Pitt

1	SPORTS GROUND	10	STRUCTURAL YARD ENTRANCE
2	WESTON ISLAND WORKS	11	WORKS CANTEEN
3	WESTON ISLAND RECEIVING OF	12	CHANCELLOR'S YARD
4	AVALON WORKS	13	NEWARK WORKS
5	TRAINING SCHOOL	14	NEWARK OFFICES
6	PERSONNEL DEPARTMENT	15	NEWARK RECEIVING OFFICE
7	VICTORIA WORKS	16	OAK STREET OFFICES
8	LONGMEAD STREET STORES	17	PUBLICITY DEPT
9	VICTORIA RECEIVING OFFICE		

Above: Plan showing the layout of Stothert premises in Bath in 1980.

Left: A 15-ton steam travelling crane at Newark Foundry, 1913. This crane, made for the London & South Western Railway, shows another view of the Newark Yard. A system of standard gauge railway and turntables operated in this yard, at the rear of the works. Behind the crane is the iron and brass foundry and the crane has been 'wheeled out' from the erecting shops, where it would have been assembled, to the right.

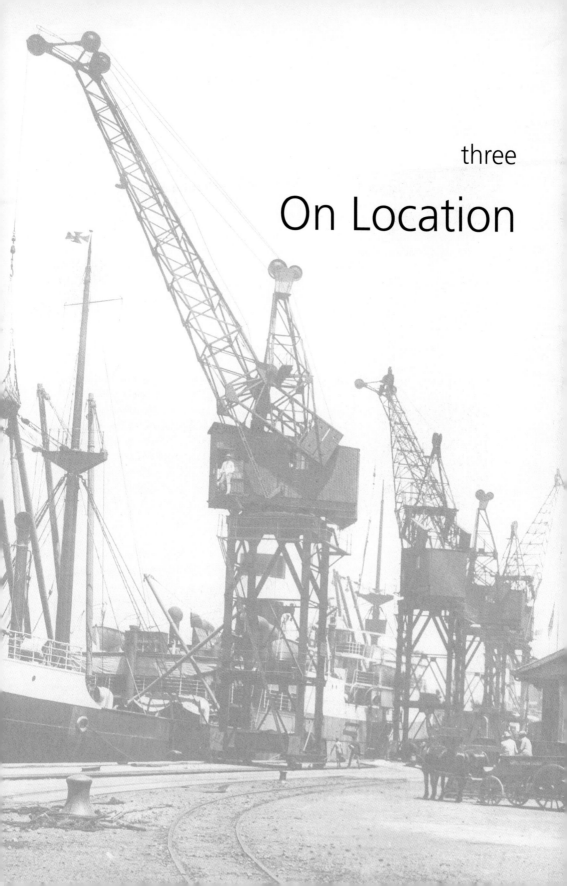

three

On Location

In the late eighteenth century George Stothert, although acting as an agent for the Coalbrookdale Co. and retailing their products, was also involved in the organisation of erecting the larger structural items being supplied. When in 1800 two bridges were erected in Sydney Gardens, a representative from Shropshire was present but Stothert would have been responsible. It is possible that Samuel Bamford's woollen mill at Twerton was supplied with specialised machinery by Stothert as early as the late 1790s. Even after the company became a manufacturing one with the opening of the Horse Street foundry in 1815, practically all the supply and installation jobs were in the Bath area and this continued into the 1820s. In 1823 fairly large treadmill installations were equipped at Shepton Mallet and Taunton 'Houses of Correction'. The involvement with the building of the Great Western Railway tunnel at Box and the exhibition successes at the Great Exhibition in 1851 had established the company as a dynamic engineering concern with specialisms in cargo handling and pumping. By the time the new Newark Foundry had opened in 1857, the growth of the railway network and the upgrading of docks and ports had created a demand for cargo-handling equipment of all kinds and, as the market for cranes developed, Stothert & Pitt supplied cranes all over the United Kingdom. The provision of water supply equipment for installation with Brunel's prefabricated hospitals for the Crimean War drew the government's attention to the company, and in 1860 it was reported that:

We gathered that the firm has been employed by the Great Western, South Wales, Bristol and Exeter, South Devon, Oxford and Birmingham, Cornwall, Melbourne and Hobson's Bay, Grand Trunk of Canada and other railway lines to supply Cranes, (water) Tanks, Water Cranes, Pumps, Gas Apparatus, Weighing Engines, Turntables and all kinds of railway fittings; by the Government and the Trinity House for Cranes and Lighthouse requirements – that they have supplied fixed or travelling cranes to the Docks and Harbours at Dover, Hull, Southampton, Swansea, Cardiff, Aberdeen, Jersey, Australia, etc. Gas Works to the towns of Reading, Exeter, Bradford, Bridgwater, Tenby and the principal towns of the West of England; also heavy contracts in the island of Cuba and many sets for private establishments; Heating, Cooking and Washing Apparatus at the Devon and Warwick County Asylums and several Baths and Laundries and Soup Kitchens; Water Works and Pumping Engines at Swansea, Cheltenham, Chepstow, Teignmouth, Bath, Clifton and other provincial towns.

The development of concrete mixing and delivering machinery, as well as the creation of the series of heavy load-lifting 'Block Setting' cranes for installing concrete blocks in artificial harbours, created a new international market for the company. Titan class cranes were supplied to all the port authorities looking to not only improve their facilities but to extend and enlarge them. So large were these cranes that, when their job of creating the harbours was complete, many remained for years before being scrapped. Around the same time, the development of the company's first electric crane (1893) and the revolutionary Toplis level luffing mechanism (1912) guaranteed the continuing international demand for cranes.

International markets for other products such as pumping and civil engineering equipment were aided by the patronage of the Admiralty, Crown Agents for the Colonies and governments both in the United Kingdom and abroad.

After the company closed its manufacturing operation in 1989, the new owners of the drawing office records were guaranteed business owing to the large number of requests from abroad for details of replacement parts for the cranes and pumping equipment installed decades earlier. The manufacture of the requested spare parts from Stothert & Pitt's extensive drawings is now subcontracted to other manufacturers.

Above: A 1½-ton steam dredging crane on the Kennet & Avon Canal, near Bathampton, 1925. The Great Western Railway took over the Kennet & Avon Canal in 1852 and continued maintaining it, dredging with light, boat-mounted cranes like this until 1948. The width of the pontoon was restricted to allow passage through canal locks.

Right: Electric roof crane at Bristol Docks, *c.*1905. Mounted on the secure dockside storage sheds at Bristol City Docks, this fixed crane could raise loads up to 2 tons. This is an early electric installation, the first Stothert electric crane had been completed only twelve years earlier.

Above: Electric cargo handling crane at Southampton. The first electric crane made by Stothert & Pitt was supplied to Southampton Docks in 1893 and is seen here shortly after installation.

Right: A 1½-ton 'semi-portal' dock crane, Liverpool Docks, 1915. These light load cranes were some of the earliest cranes to be fitted with the Toplis level luffing equipment, invented just a few years earlier. This improvement allowed the load to be kept level regardless of the angle of the crane jib. These cranes were known as 'semi-portal cranes': although they arched over the docks and could travel over obstacles, they were mounted to the storage sheds on the dockside.

Titan harbour block-setting crane, South Shields, 1901. The Titan class of cranes, for construction of artificial harbours, were supplied from the 1880s. Each had an enormous radius of 100 feet and most could lift concrete blocks up to 41 tons in weight. Rail mounted, they advanced on the harbour their 'block setting' had created.

Above: Dockside cranes at Durban Harbour South Africa, 1923. These 4-ton electric travelling cranes were fitted with the Toplis level luffing equipment for cargo handling.

Left: Ten electric cranes at Santos Docks, Brazil, 1928. In 1927 this set of 6-ton loading cranes were supplied to the newly modernised dock and are seen fitted with 'grabs' for mineral handling.

Left: Cranes at Gyproc Co.'s wharf on the Clyde River, Glasgow. These two 5-ton cranes – loading railway wagons with gypsum products – were supplied in the 1930s.

Below: Cantilever transporter cranes at Ferrybridge Power Station, Yorkshire, 1962. Where it was only necessary to unload or load 'on the level' over relatively short distances, cantilever transporters such as these could be used. They are shown here trans-shipping coal from barges.

Left: DD2 cranes at Grangemouth. These cranes were supplied as part of a large installation of DD2 dockside cranes to Grangemouth Docks and were capable of lifting up to 6 tons.

Below: DD2 dockside cranes at Royal Portbury Docks, Bristol, 1970. The tubular steel construction of this award-winning crane made erection more easily accomplished. A most successful and aesthetically pleasing design which, despite docks having closed to commercial traffic, has ensured that examples have survived as dramatic landmarks in new developments such as Docklands in London.

Opposite above: Hammerhead Dock Crane, Rosyth Dockyard, 1965. Some of the largest cranes capable of the heaviest loads were supplied to Royal Navy construction yards. This crane, supplied in the 1960s, could lift loads up to 120 tons and had a maximum reach of 180ft.

Below: Container transporter at the Trafford Park basin of the Manchester Ship Canal. By 1968 when this installation was supplied, the transport of cargo in standard containers had begun to replace smaller 'loads'. These cranes were supplied during a modernisation of the canal basin and were able to lift up to 25 tons in weight.

Semi-portal DD2 cranes at Liverpool Docks, c.1965. Shown here is the 1960s version of the semi-portal crane type. The lifting capacity was 5 tons.

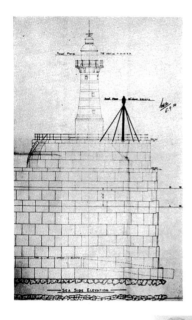

Lighthouse for Peterhead Harbour, 1913. The surprising diversity of Stothert & Pitt's products is demonstrated in this illustration of an iron lighthouse, constructed in 1913 for the South Breakwater Head, Peterhead, Aberdeen.

Toplis electric levelling davits for lifeboats on SS *Arundel Castle* at Southampton Docks, *c.*1920. The Toplis self-levelling arrangement was ideally suited for the lowering of lifeboats or side-boats and keeping them level and stable.

Multi-bucket excavator at Clock House Brick & Tile Co., Capel. Soil excavators had been available from the company since the late nineteenth century. This petrol-engined example, digging clay for bricks, was supplied in the 1930s.

Belt-conveyor installation at opencast workings, *c.*1930. Large installation such as this, with a design based on crane-making practice, were supplied for earthmoving or construction projects. These installations were designed specially for customers' requirements.

Travelling transporter crane, British Iron & Steel Co. Ltd, Cardiff, *c.*1935. This rail-mounted crane was used to move materials to the blast furnaces. The massive towers of the steel-making plant may be seen in the background.

Above: Mineral handling plant, Port of Liverpool Authority, *c.*1935. This fixed conveyor for minerals is pivoted, allowing movement in an arc.

Right: Vibroll compactor, 1960. The Vibroll self-propelled compactor used heavy weight and the vibration of the rollers to flatten surfaces. The ease of using this equipment is clearly shown in this picture of domestic bliss with a couple laying their new drive.

Extension work to Bath Bus Depot, *c.*1963. At work in the early 1960s on extensions to Bath Bus Depot at Kensington, London Road, Bath, are the 7T Hydraulic Tilting Drum Concrete Mixer and Vibroll 28W lightweight vibrating roller.

Concrete mixer at Kilcobben, Cornwall, 1968. Stothert & Pitt concrete mixers at work on the new RNLI Station, Kilcobben Cove, Lizard, Cornwall.

Right: Concrete mixing and distributing tower, Metur Dam, Madras, India, 1928. This huge structure, one of two towers erected by Stothert & Pitt, supplied concrete for the construction of an enormous dam on the River Cauvery in India. Each rail-mounted tower weighed 2,000 tons and was capable of delivering 300 tons of concrete each hour.

Below: The Metur Dam under construction showing concrete placing tower, 1928.

Right: Electric capstan at London Passenger Transport Board Yard. Although principally designed for dockside requirements, capstans could be supplied to assist in simple shunting or securing operation in railway goods yards. Here shunting of wagons is shown to be possible by hand, assisted by the electric capstan.

Below: Crushing, screening and discharging plant, St Keverns, Cornwall. The entire installation for this granite quarry was installed by Stothert & Pitt in 1966. The contract included the equipment to discharge the processed mineral aboard ship. In 1966 this installation cost £600,000.

four

People and Production

The original ironmongery business George Stothert acquired in 1785 also included a team of metal smiths and plane makers. In addition, a certain amount of installation or repair would have been contracted out to local workmen. With the opening of the iron foundry in 1815, however, the workforce and the scale of work would have changed. A workforce of founders, pattern makers and machine tool operators would have been required as well as workers for installing the manufactures.

By the time the Newark Foundry had opened on the Lower Bristol Road site in 1857, the scale and range of work undertaken demanded the new spacious accommodation and the fitting-out with large precision machine tools as well as facilities for iron, brass and alloy founding and, increasingly, administration staff. It is not known how many staff Stothert & Pitt were employing by this time, but it was probably in the hundreds. Precision engineering was at a premium with the development of fixed and travelling cranes and pumping equipment, and, with the installation of more and more equipment, the chances of serious injury must have increased. In December 1891 the *Bath & Cheltenham Gazette* reported:

A Shocking Accident
On Monday afternoon a young man named John Bryant in the employ of Messrs Stothert & Pitt, iron founders [of] Lower Bristol Road and residing at 8 Kingsmead Square was attempting to put a strap on some machinery, that was in motion, when his right arm was caught by the wheels and the limb was completely torn off at the elbow joint. The sufferer was taken to the United Hospital where he had to submit to a further amputation in order to stop the haemorrhaging.

By 1900, with the Victoria Works opened and an increase in international business for cargo-handling equipment, the workforce numbered around 1,000, and by the end of the Second World War this had increased to nearly 2,000.

By this point the metal-casting and machining had been augmented by the need for fabricating large installations such as the concrete mixing plants and block-setting cranes in the huge erecting sheds of the Victoria Works.

The purchase of the company by the Hollis Group, and the resulting rationalisation, reduced the workforce from around 800, and by 1989 only 580 people were still employed by the company.

Fitting Shop at Torrance & Sons Works at Bitton, 1925. Although Stothert & Pitt had acquired Torrance & Sons five years before this picture was taken, the name was retained for this branch of the operation until after the Second World War. Here workmen are assembling roller mills.

Boiler assembly at Victoria Works, 1928. Steam-powered cranes required their own steam-raising equipment and very large boilers were needed for the largest of cargo-handling cranes. In time, with the replacement of steam power by diesel and electric power, these workshops were expanded and used for general structural work.

Work in Structural Department, Victoria Works, 1980. The boiler shops became the structural shops. Here, towards the end of the company's history, a crane jib is seen under construction.

Welding steel bars, Victoria Works, 1980. Fabrication of structural elements for cranes or contractors equipment, often to customers' specific instruction, involved welding of steel components.

Vertical lathe, Victoria Works, 1977. Very large pieces of equipment were employed in machining and many of the machine tools sold in 1989, when the company closed, were less than ten years old.

Gear-cutting machinery, Victoria Works, 1976.

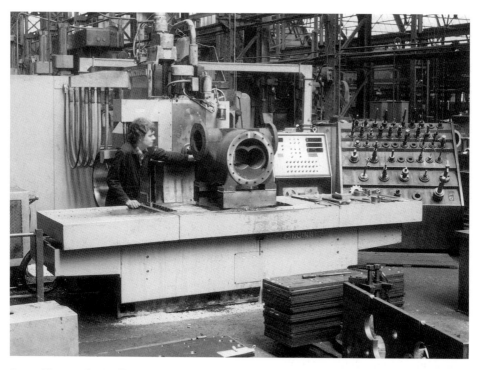

Pump Shop marking off table, 1980. The main 'block' of a pump is shown awaiting the insertion of the revolving internal parts.

Large lathe, Victoria Works, 1977. Of great length, this machining lathe is equipped with a moving shield to allow the machining to be examined at close quarters with safety.

Profile burning machine, Victoria Works, 1980. The precision-cutting of larger structural components for cranes, etc., could be undertaken with this piece of equipment.

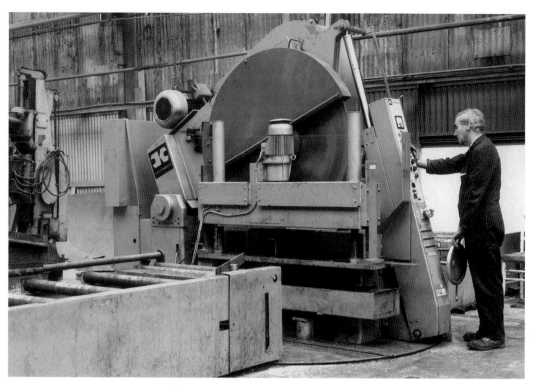

Operating a high-speed saw, Victoria Works, 1977. This fearsome machine could cut precise sections of steel, and the cutting wheel is in the process of being gently lowered onto the work to be cut.

Welding in Structural Shop, Victoria Works, 1980.

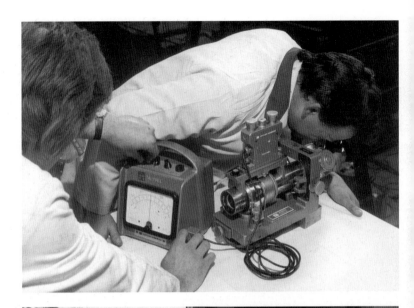

Right, top: Quality control testing, 1977.

Right, middle: Computer-controlled batching plant, 1981. Around 1980 the first computer-controlled measuring equipment was supplied by Alkon & Co. for operating with Stothert & Pitt's concrete and asphalt mixing plants. Here a 'batcher' checks that the computer has prepared the correct mixture of ingredients to produce the finest concrete.

Right, below: Testing a Vibroll compactor, 1980. Many of the company's products were tested 'in the field'. Here the double-roller Vibroll compactor is being eased onto a low truck under its own power.

Welding circular girder. This photograph shows a circular girder being assembled and welded in the structural shop. It was 31ft in diameter and weighed 30 tons. This girder was for an electric travelling crane for Silley Cox & Co. Ltd, Falmouth, ship repairers and engineers.

Left: Computer-controlled gear cutting, Victoria Works, 1980. This early – and cumbersome – computer by General Electric was used to guarantee a precise finish to the cutting of gear wheels.

Workmen of the Pattern Shop, 1898. The workmen of the pattern shop include the foreman, Mr O.S. Grace, and Sid Shore, who is fourth from the left. Next to him is Mr A. Andrews. Also in the picture are Messrs J. Bailey and T. Loran, and in the front the bowler-hatted gentleman holding the spur wheel is Mr R.H. Hope, JP.

Workmen outside Newark Foundry, 1915. Some workers have been named: H. Lintern, C. Maylott, W. Hinton, J. Ayres, T. Paul, W. Stokes, E. Howell, W. Worle, H. Flower, G. Howell, F. Underbill, J. Green, A. Morfatt, A. Ringham, G. Rogers, A. Cashnella, A. Hatherell, W. Poole, W. Hawkins, H. Frankham, J. Hardick, E. Bowler, W. Hiskins, L. White, A. Lodder, A. Freegard, R. Collett, A. Tucker, F. Miles, F. Hinge, W. Reeves and E. Maylott.

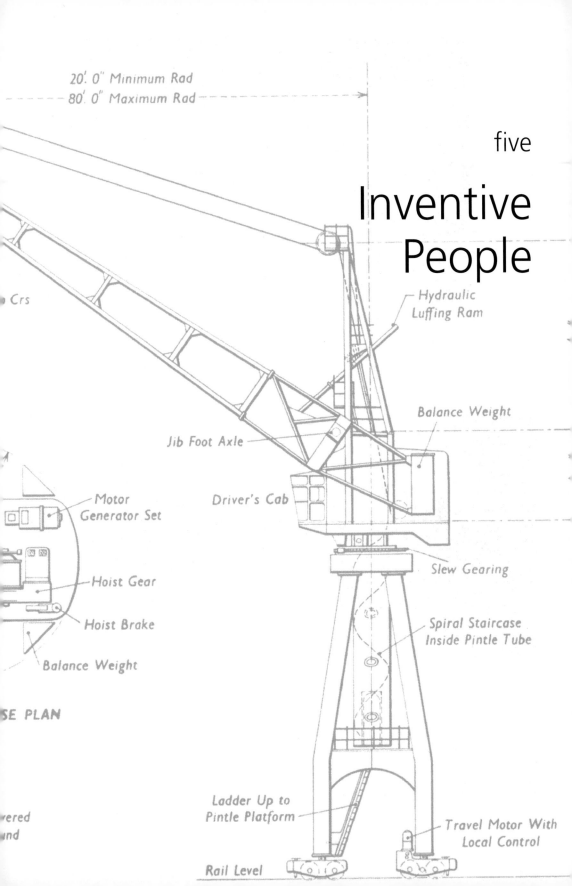

20'. 0" Minimum Rad
80'. 0" Maximum Rad

five

Inventive People

Hydraulic
Luffing Ram

Crs

Balance Weight

Jib Foot Axle

Driver's Cab

Motor
Generator Set

Slew Gearing

Hoist Gear

Spiral Staircase
Inside Pintle Tube

Hoist Brake

Balance Weight

SE PLAN

Ladder Up to
Pintle Platform

vered
and

Travel Motor With
Local Control

Rail Level

Two people were outstanding in the major growth years of the company other, of course, than family members and partners. These were Claude Toplis, at one time their chief engineer, and Ernest Feuerheerd who, whilst not a member of the company, greatly influenced its fortunes in the 1920s with his inventive pump design.

The name of Toplis is world famous in the crane and lifting industries; his invention of the 'level luffing' system was a breakthrough in speeding-up the loading and unloading of ships. Luffing, or derricking, is the process of moving the jib of a crane in or out. Normally the load would be raised or lowered due to the varying height of the jib, but level luffing kept the load at the same level. This meant that the crane driver would have fewer operations to carry out during loading or unloading, thus speeding-up the process and greatly pleasing dockers who were on piecework and saw increased earnings. The Toplis system was also used on ships for lifeboat davits.

Toplis seems to have been prolific in his inventiveness, having applied for no less than twelve patents between 1905 and 1930 – his level luffing patent was granted in 1912. His other patents related to cranes, electrical machines and even pianos.

Whilst the Toplis name is well known, Feuerheerd never acquired the same publicity and seems to be little known. Ernest Feuerheerd was of German descent, though it was three generations since the family had been German; his father was a naturalised Englishman. Feuerheerd worked at the invention of his Rotary Lobe Pump between 1900 and 1919 when it was patented. He then set up a company – Feuerheerd Rotors Ltd – to market his product. Sales were not good due to anti-German sentiments following the First World War – there were even advertisements exhorting people not to buy these 'Junkers' products when perfectly good English ones were available, this overlooking the fact of course that Feuerheerd was native-born and had served in the British Army during the war. This had a disastrous result for the business and things looked bleak until Stothert & Pitt started to manufacture the pump under license, when sales began to boom. The pump was extremely useful, as it could handle a large range of viscosities and was reversible.

Feuerheerd's enterprises continued but he never seems to have prospered, though he did eventually have a dubious claim to fame having gone into partnership with John George Haigh – the 'acid bath murderer'. It was not until after Feuerheerd's death in 1948 that the murders came to light, but this is possibly one of the reasons for his name having faded into obscurity.

Another inventive mind from the later years of the company was Norman Kerridge, who retired as Director of Research & Development in 1971 and was credited with the design of the DD2 dockside crane – a streamlined version of the previously lattice-built cranes. He was also responsible for the design of the world's first commercially viable vibrating roller after the Second World War. Kerridge, like Toplis, was a prolific generator of patent applications.

The Drawing Office team of 1920. From left to right, back row: R. Hodges (later senior executive of Spencers, Melksham); E. Bellringer (later senior executive for Construction Equipment Division); J. Rainey; H. Nicholls; A.W. Mellor; C.A. Brooks; R. Caple; R. Clark; ? Gilmore (brother of K. Gilmore, another S&P man then retired); H. Dyke; J. Auld; E.L. Hocking; -?-; E. Martin (tracer); H.N. William (DO Store); E.A. William (photographer and chief of tracers). Middle row: W. Williams (father of H.N. Williams in back row); H. Poole; A. Grist; W.W. Padfield; A.O. Helps; E. Blake; H.J. Tanner; C.T. Pearce; P.A. Arbenz; E. Evans (chief draughtsman and steam crane expert); H.A. Rushton; J.L. Wellings (chief of Concrete Mixer Section); T.A. Briggs (chief estimator, Cranes); G.H. Wiltshire; N.D. Hope; ? Wheeler; C. Lawrence; ? Hudson; W. Angraves; R. Everard; W. Jones. Sitting: H. Davis (typist); Miss Sekington; Miss Rushton; two sisters, the Misses Sheppard; a tracer from Bristol; Mr Toplis (chief engineer and inventor of the Toplis level luffing system); A.O. Day (later to become a colonel in the Royal Engineers); Miss Baker; Miss W. Grove; Miss Lloyd (Melksham); Miss M. Hand; Miss O. Coleman; and the office boy (name unknown).

A 5-ton cargo-handling crane at Southampton Docks fitted with the Toplis level luffing arrangement, c.1935. In this silhouette view the rope level luffing arrangement on the crane jib is well illustrated. One of the two Union Castle four-stackers is visible.

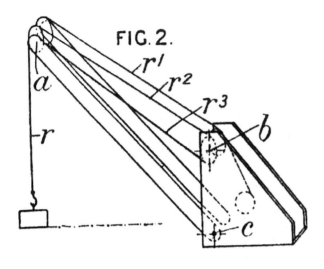

Sketch of Toplis level luffing mechanism, 1912.

Above: Ernest Feuerheerd, *c.*1927.

Left: Advertisement for Feuerheerd Pump Co., 1920. Feuerheerd had set up his own manufacturing operation but Stothert & Pitt soon began making his pumps under licence.

Norman Kerridge, Director of the
Research & Development
Department in 1959.

An array of DD2 tubular steel
cranes.

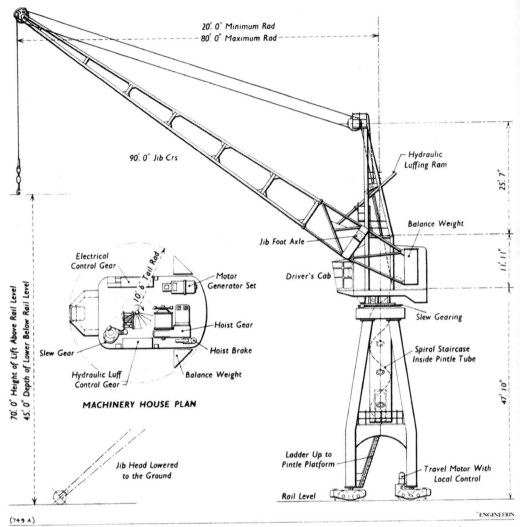

Fig. 5   Drawing of the DD2 5 ton model as being supplied to the Port of London

Particulars of Typical DD2 Type Cranes

	5 tons at 80 ft. radius model	3 tons at 65 ft. radius model
Jib centres	90 ft.	72 ft.
Tail and clearance radius ...	10 ft. 6 in.	10 ft. 6 in.
Rail centres	13 ft. 6 in.	10 ft. 1 in.
Cab floor height	47 ft. 10 in.	39 ft. 9 in.
Jib foot axle height ...	59 ft. 9 in.	51 ft. 8 in.
Hoist motor h.p.	60	45
Hoist maximum speed ...	0 to 5 tons 425 to 150 ft./min.	0 to 3 tons 425 to 180 ft./min.
Slew speed	1¼ r.p.m.	1½ r.p.m.
Luffing speed	0 to 160 ft./min.	0 to 160 ft./min.
Travel speed	50 ft./min.	50 ft./min.

A line drawing and details of the DD2 crane from the *Stothert and Pitt Magazine*, 1959.

six

Products

When George Stothert opened his iron foundry in Horse Street in 1815, the product range was based on that of his ironmongery suppliers and, in particular, those of the Coalbrookdale Co. This included domestic iron fittings, notably cooking ranges, bread ovens, fireplaces, garden gates and small-scale constructional materials. Woodworking planes were also manufactured. The original foundry was quite small but capable of the casting of the two small bridges which span the Kennet & Avon Canal at Widcombe Locks in Bath. By the early 1820s treadmills were being supplied to the 'houses of correction' at Shepton Mallet and Taunton, and the firm seems to have developed them as a speciality, supplying Bath Gaol with one in 1867. In 1827 George Stothert patented a plough for manufacture. The firm were also agents for other products, including Henry Marriott's portable weighing machine and corn mill, and Finlayson's agricultural harrow. Iron drain gratings were supplied to Bath Abbey, and in the 1830s water pumps were supplied to the Kennet & Avon Canal Co. for their Crofton Pumping Station and Bath Corporation for its hot water springs.

One of the first recorded cranes manufactured by the company appeared on display at the Great Exhibition of 1851, rather oddly marked 'Stothert' despite the fact that the company had by this time become Stothert & Pitt, and in the 1860s a patented leather dressing machine was put into production. In 1860 a visitor to the newly opened Newark Foundry described the scene:

the work in hand at the time of our visit consisted of a 50 foot waterwheel for a mill to manufacture paper from straw, cranes of all sorts and sizes from 20 tons downwards, a weighing engine of 25 tons power, an iron lighthouses intended for New Zealand, a very powerful brick and pipe machine ... several compact steam engines of inexpensive construction ... steam boilers and an iron roof 480 feet in length for the Sutton Copper Works at Liverpool.

In the same year the firm was reported to have supplied railway companies, Trinity House, docks and harbours from Australia to Hull, gas works, laundries, pumping stations, soup kitchens, waterworks and asylums.

An existing 1885 catalogue of 316 pages lists 1,050 items, including horsehair envelopes (for oil mills), punching machinery and steam yachts and tugs. Many of the items advertised, for example tea urns, would have been made elsewhere and marked with the Stothert name. As early as the 1860s the company was also supplying concrete mixing and distributing equipment for harbour-building and other large-scale constructional projects.

The largest of cranes, for heavy constructional work, first appear in the 1890s, by which time the Victoria Works had opened with adequate space for these massive machines. Block-setting cranes known as Titan cranes could lift up to 50 tons, whilst the Goliath cranes could lift up to 250 tons. In 1912 Claude Toplis's level luffing patent for level handling made Stothert cranes technologically superior to all other lifting gear.

The supply of deck machinery for merchant and fighting navies complemented the dockside installations, and during the First World War the British Royal Navy ordered cranes and minesweeping equipment. An experimental tank, the *Pedrail*, was supplied to the Trench Warfare Department of the Army.

Although pumping equipment had been supplied from the earliest days, in the 1920s Ernest Feuerheerd's rotary pumps went into production, and in 1936 positive displacement screw pumps first appeared. Concrete mixers and other building equipment were supplied, and in 1928 a concrete mixing and distributing installation 286ft high helped in the construction of the Metur Dam at Madras.

By the 1930s the company's name had become synonymous with dockside cranes the world over, and, as docks and harbours developed with new methods of cargo handling, so the company's products changed. After the Second World War, the development of a German idea for road-rolling using vibrating plates resulted in the Vibroll range of rollers, the first exploitation for this purpose. Designer Norman Kerridge, who was responsible for the Vibroll system, also developed the revolutionary DD2 cranes in 1959.

In 1968 and 1970 the company received design awards for its cranes, and the range of road-rolling, pump making, constructional ironwork and crane manufacture continued until the firm's closure in 1989.

Tea urn, c.1885. These tea urns are advertised in the surviving 1885 catalogue of the company's products. It is probable that they were not manufactured by the company but supplied by them.

MESSENT'S PATENT CONCRETE MIXER.

Messent's concrete and mortar mixer, 1885. By the 1880s Stothert & Pitt were manufacturing these hand-operated, rail-mounted mortar mixers, patented in the 1860s by P.J. Messent.

Victoria cement mixer, 1933. Cement mixing equipment had been supplied by the company since the late nineteenth century. By the 1930s the popular Victoria mixers were available in a variety of sizes.

Early mortar mixing machine, *c*.1908. This early hand-operated mortar mixer was described as 'in constant use in buildings where steam power is not available and found thoroughly efficient for mixing hair and other mortar, concrete, cement, etc. The machine, with two men, will mix 10 to 12 cubic yards per day.'

Transportable concrete batching plant, 1980. Batching plant for the accurate measuring and mixing of ingredients for concrete or asphalt was made to be easily dismantled and moved. Known as Transbatch plant, this equipment was composed of modular components.

Containerised Concrete Plant (CCP), 1980. Here part of a concrete batching plant is shown containerised for ease of shipment as a lorry load or ship cargo.

Macadam crushing, screening and coating plant, 1955. Whole installations for the preparation of road-making materials were supplied by the 1930s, including this self-contained plant erected for the Whitwick Granite Co. Ltd. A rail connection allowed the processed materials to be delivered to wagons.

Above: 'Dustless' asphalt-making installation, 1972. This asphalt-making equipment made by Stothert & Pitt under licence from the Wibau Co. of Germany is shown here being tested at Stothert's own testing installation at Chipping Sodbury, Gloucestershire. Subsequently it was supplied to Amalgamated Quarries for their works at Stirling.

Right: Travelling Scraper at Newark Foundry, *c.*1930. This equipment, derived from the light multi-bucket excavator, was supplied to the Ammonia & Nitrates Co. and could move 100 tons per hour.

FA820

Above: Edge Grinding Mill at Torrance & Sons, *c.*1930. The firm of Torrance & Sons had been acquired by Stothert & Pitt in 1919 and had specialised in the manufacture of high performance grinding mills for paint, etc. Later, roller mills were also made.

Left: Rotary Lube Pump for Shell Petroleum, 1950. The rotary lube pumps, which Stothert & Pitt had begun manufacturing to Feuerheerd's designs in the 1920s, were progressively improved and developed. This fuel oil pump, showing the revolving pump assembly, was driven electrically and its speed could be adjusted through a gearbox.

Portable Screw Displacement Pump. One of several portable screw displacement-type pumping units, supplied to the Admiralty for handling oils of medium viscosity. Screw displacement pumps began to be manufactured in the 1920s and used revolving screw shafts within the assembly to move the liquids through the pump.

Screw Displacement Petrol Pump, 1943. This petrol engine-driven mobile screw displacement pump has an output of 300gpm of petroleum spirit when running at a speed of 1,000rpm. A total of 149 pumping units were constructed for the Ministry of Supply during the Second World War.

Ship's windlass at Victoria Works, *c*.1935. Deck machinery of all kinds was supplied, including capstans, windlasses, winches and deck cranes. Windlasses of this standard design could haul up to 50 tons.

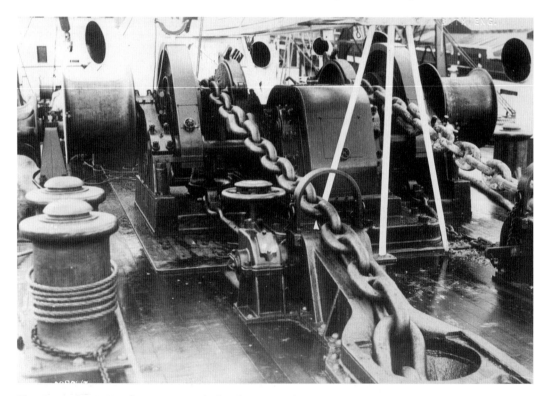

Electric winch, *c*.1950. These items, supplied with magnetic brakes and occupying a minimum of deck space, could be used for hoisting anchors, as in this case, or general winding and lifting duties aboard ship.

Right: Bucket reclaimer, 1967. Under an agreement made with ERMI of Paris in 1967, the company had a licence to manufacture this giant bucket wheel reclaimer for sale in the UK, Australia and New Zealand. The agreement embraced a standard range of rail-mounted machines for capacities from 40 tons per hour up to 7,000 tons per hour, in addition to special-purpose equipment. The commodities handled included coal, iron ore and other minerals.

Below: Fairbairn steam crane at Bristol City Docks. This crane, built to a patented design by William Fairbairn, was installed by Stothert & Pitt in 1878. This crane, which could lift up to 35-ton loads, has not only survived but been restored to full working order by Bristol City Museums.

Above: Goliath-type heavy lifting crane at Shoeburyness, 1922. This very heavy-duty crane had a lifting capacity of 250 tons and was supplied to the Ministry of Munitions' Experimental Testing Facility at Shoeburyness. Seen here lifting the barrels of warships, the rail-mounted crane cost £14,000 in 1922.

Left: Snatch Block from Goliath crane, 1922. This photograph shows the assembly which allowed very heavy weights to be born by the lifting gear of the Goliath class crane supplied to the Shoeburyness Testing Facility of the Ministry of Munitions.

Titan-type block-setting crane for Peterhead Harbour, 1891. This Titan-type crane was the largest ever built; it weighed 576 tons and was supplied for the building of the Peterhead Harbour of Refuge. The crane had a lifting capacity of 50 tons and, on site, an auxiliary Goliath-type crane supplied the Titan with 50-ton concrete blocks. This crane was finally scrapped in 1956.

Travelling Goliath-type crane at Brighton goods yard, 1940. This traverser crane is movable, running on its own set of rails, and was supplied to the Southern Railway. Its lifting capacity was 12 tons.

Left: Deck crane on RMS *Olympic*, 1911. Deck cranes like this, used to lower smaller loads aboard passenger liners, could lift up to 50cwt. Cranes identical to this were supplied to Harland & Wolff for the *Olympic*'s sister ship, RMS *Titanic*.

Right: DD2 tubular jib crane at Newark Foundry, 1959. The DD2 crane, designed by Norman Kerridge in the late 1950s, proved to be one of the most popular dockside cranes. Here an early example demonstrates its capabilities to an invited audience at the Newark Foundry on 5 July 1959. An earlier example of a cargo-handling crane stands by.

Right: Steam gantry cranes at Victoria Works, 1921. Rail-mounted steam-handling cranes were still being supplied by Stothert & Pitt in the late 1930s. These rather utilitarian examples, lifting up to 30cwt, were ordered by Wilson, Sons & Co.

Below: Steam breakdown crane, 1913. Stothert & Pitt supplied a great number of products through the Crown Agents for Colonies who equipped government-administered or sponsored commercial activity. Self-propelled railway breakdown cranes like these were supplied by the hundred to foreign and colonial governments.

Right: Extending/traversing container crane, 1969. The requirements of the container cargo trade prompted the development of cantilever traversers for the unloading of the standard size containers such as this. The traverser could be extended 40ft out over the dock side.

Left: Offshore crane for oil and gas platforms, *c.*1980. The requirements of the offshore oil and gas extraction industry in the 1970s included great strength whilst taking up very little deck space aboard an extraction platform. These cranes proved very successful and were still being made when the company closed in 1989.

Vibroll compactor. After the Second World War, German experimental technology was employed in a range of surface compacting machinery. The rollers compressed the surface below with weight but also vibrated to compact the surface more efficiently. By 1985 forty-five countries were distributing Vibroll products. The second view shows a larger, steerable Vibroll compactor at work on the Lower Bristol Road in Bath. Very large applications of the vibrating roller equipment were available used in the construction of motorways or airport runways.

Loading the *Queen Mary* at Southampton Docks, 1936. Southampton Docks had ordered a large number of Stothert cranes during the 1930s in part to service the new and large transatlantic liners such as the *Queen Mary*. In this view a series of 5-ton cargo cranes are shown.

seven

War

As early as the late 1700s Stothert was supplying weaponry, when the operation advertised that:

the members of the Bath Armed Association may be supplied with warranted firelocks at £2 each at Stothert & Co warehouse at No.15 Northgate Street, likewise pistols and swords and belts and cartouche boxes

Although contracts had been undertaken by the company installing equipment for hoisting explosive shells at coastal defence batteries and other work for the Navy, it was not until the First World War that specially commissioned and designed goods were supplied. Admiralty contracts had long been undertaken, but now paravanes and minesweeping equipment were designed and large cargo cranes were supplied to dockyards. Ammunition was also produced, with large orders of 18-pound shells for the Ministry of Munitions.

Experimental tanks were designed and manufactured and in use in France by 1916.

With the approach to the Second World War the company was supplying deck equipment to the Navy, notably seaplane-lifting cranes to cruisers and battleships. At the outbreak of war the company was also supplying graving dock cranes at Gibraltar, Singapore and Sydney, mobile fuel pumps, the aforementioned minesweeping equipment and Human Torpedoes. The Human Torpedoes were two-man mini submarines which, having been launched from warships, would drop a large explosive warhead below enemy warships at anchor. Production began in circumstances of great secrecy in 1942 at the Victoria Works, and the 'Chariots' as they were known were highly successful in destroying Italian shipping in the Mediterranean Sea.

For the Ministry of Supply, the company supplied tank equipment, including gun mountings and fighting compartments. Crusader tanks were supplied with such equipment. A large number of 6-ton cranes were supplied for rapid erection after the D-Day landing and the powder mills manufactured by the associated company of Torrance & Sons were used to grind high explosives. Concrete and tarmac-making and distributing machinery was supplied to the Air Ministry for the construction of new airfields.

In the bombing of Bath in April 1942 the Stothert Works was regarded as a military target, and bomb aimers using the River Avon to locate riverside industry were successful in bombing the factories.

New shell workshop, 1944. As well as manufacturing equipment to government contracts, ammunition shells were manufactured in new purpose-built facilities at the Victoria Works.

No.1 Military Port, Gareloch, Scotland. After the fall of France in 1940, British south and east coast ports were virtually closed to large ocean-going ships. Two ports were, therefore, built on the west coast of Scotland. In this view, some of their cranes are shown installed at No.1 Military Port, which is in the Gareloch off the Clyde.

Above: Military Port at Loch Ryan, Stranraer. Stothert & Pitt supplied cranes and other equipment for this military storage and trans-shipping base.

Left: Electric capstans for Admiralty – 576 electrically-driven capstans of the type shown in this illustration were supplied to the Admiralty. These capstans were designed for a maximum pull of 10 tons at a speed of 8ft per minute in slow gear or 1 ton at 80ft per minute in fast gear.

Opposite above: Seaplane cranes aboard cruiser HMS *Superb*, 1944. Here is shown an example of the twenty-two seaplane cranes the company built for the Fiji class of cruisers of the Royal Navy during the Second World War. The lifting capacity of each crane was 6 to 7 tons. More than 130 cranes of various types – built to Admiralty specifications – were produced by Stothert & Pitt during the war and were fitted to many different vessels, from minelayers to aircraft carriers.

Opposite below: Seaplane crane aboard ship, 1940. Lifting a reconnaissance seaplane from the water outside Plymouth's naval dockyard.

A 'Human Torpedo' being launched. In January 1943, a number of these craft – known also as 'Chariots' – were used with great success when they sank the new Italian cruiser *Ulpio Traiano*, and severely damaged an 8,500-ton transport vessel in Palermo Harbour. These craft were designed and built by Stothert & Pitt in close collaboration with the Admiralty.

A Human Torpedo underway. Each of the crew of the Human Torpedo wears several layers of thick woollen clothing and a 40lb diving suit with visor. When these craft had reached their target, the warhead was detached and attached to the warship.

Landing ship. Many ships of this type were fitted with bow door operating gear, which was designed and manufactured in the works. The ramp operating gear was also made to Stothert & Pitt design.

Fixed gantry cranes for handling landing craft. Fixed gantry cranes were supplied for these vessels. The illustration shows one of the runways in the 'outboard' position preparatory to discharging a landing craft.

Landing ship of the Thruster class. The cantilever landing bridge shown in this picture is in its extended position having 'landed'. The whole of the design and manufacture of this type of bridge was carried out by Stothert & Pitt. The bridge, which was designed to withstand loads up to 40 tons in weight, is shown carrying ashore a Churchill tank.

Close-up view of Stothert & Pitt bow doors on the landing ship.

Gun-mounting in a Churchill tank.

Above: The Challenger tank. The whole of the design, manufacture and supply of the 17-pounder gun-mounting, as well as the intricate turret traversing mechanism, together with the turret design for each and every Challenger tank produced was carried out by Stothert & Pitt.

Left: Super 54 Victoria portable concrete mixer. This machine, which was primarily designed for the construction of aerodrome runways, was capable of an output of 720 cubic yards of mixed materials per eight-hour day, and was the first machine of this size manufactured in this country to be fitted with a mechanically-operated loading hopper and mounted as a portable unit.

A No.16 tarmacadam plant. Ordered in large numbers by the Ministry of Supply for the War Office, machines of the type illustrated were used extensively for the construction of vital supply roads for the fighting services. This machine was capable of producing 140 tons of tar or bituminous macadam per eight-hour day.

Conical pan mills for explosives. Supplied by the associated company Torrance & Sons Ltd, Bitton, Gloucestershire, this illustration shows part of a batch of seventy-two conical pan mills in course of construction. These mills were specially designed for the manufacture of high explosives and were installed in many of HM ordnance factories.

Dockside cranes for Ministry of Supply, c.1944. The Ministry of Supply commissioned 360
dockside cranes of this design. All these cranes were built by the company in eighteen months
and at one point the rate of production was as high as ten per week. Two major considerations
influenced the design. Firstly, the cranes had to be split up into units to permit manufacture 'on
site' by different contractors and all the units were designed for interchangeability. Secondly, the
cranes had to be capable of quick and easy erection under enemy action. It is on record that some
of these cranes were erected and put to work in twenty-seven working hours. Each crane was
designed with its own diesel generator set, thus making it independent of local electricity supply –
a very necessary feature for cranes installed at ports heavily damaged during fighting or at
temporary military ports.

eight

Exhibition

The major international exhibition held at the Crystal Palace in London in 1851 was intended as a showplace for British manufactures and invention and, along with many other Bath manufacturers, the company showed off their latest products.

Two exhibits were on show. The first exhibit was a plan by Henry Stothert 'for the removing of the sewage of London without disturbing the present arrangement of drains'. This envisaged the collection of sewage in distinct cesstanks from which steam engines would pump the sewage to stations where it could be deodorized before being sold as agricultural fertiliser.

The second exhibit was in the 'Railway and Marine Mechanism' section and was listed as:

an iron crane for a dock or wharf with improvements to the jib and in the general arrangement and proportion of the parts by Stothert, Rayno & Pitt of Newark Foundry, Bath – Improvers and Manufacturers

The crane appeared in a lithograph published after the exhibition had closed and, despite the company name, the simple word 'Stothert' appeared on the exhibit.

Later Victorian exhibitions also displayed Stothert products. In 1862 a 6-ton travelling crane was exhibited in Paris winning a gold medal, and the same style of crane won a silver medal at the Paris Universal Exhibition five years later. Further medals were won at the London International Inventors Exhibition in 1885 and at another exhibition in Paris in 1889.

In the twentieth century, exhibitions on the grand Victorian scale were replaced by exhibitions of manufactures of specific type. With the range of goods on offer, Stothert & Pitt products were on display at contractors, marine engineering, civil engineering, mechanical engineering and cargo handling exhibitions.

Stothert crane on exhibition at the Great Exhibition at Crystal Palace, 1851. Exhibited at the 'Railway and Marine Mechanism' section of the exhibition, this was noted as 'iron crane for dock or wharf'.

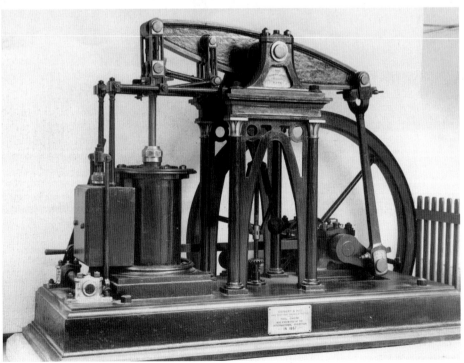

Above: Exhibition at Victoria Art Gallery, Bath, October 1945. A small display after the end of the Second World War advocating savings to consolidate the recent victory. 'Hoist your savings' indeed!

Right: Stothert & Pitt exhibits at the Festival of Britain, South Bank, 1951. In the foreground is a 7-ton warping capstan of the type supplied for British Railways' Southern Region cross-channel ferries.

Opposite above: 6-ton steam crane exhibited at the Paris Universal Exhibition, 1867. This crane won a silver medal for the company and is shown with its attached boiler and without cover for the operator.

Opposite below: Mystery engine, 1866. This steam engine stood for many years at the rear of the Newark Foundry, with an attached plate claiming it had been exhibited at the 1867 Paris Exhibition and made by Stothert & Pitt. This seems unlikely and it is most probably an apprentice's exercise or model. The engine was presented to the University of Bath in the 1970s and stands in their Mechanical Engineering Department.

Display stand at the Bath Trades Fair, 1948. In June 1948 Bath staged its own exhibition of local products. Here Stothert & Pitt exhibit a cement mixer and pumps. Other exhibitors included displays for Bath Oliver biscuits and Harbutts' Plasticine. The Mayor of Bath and company directors are shown viewing the Stothert & Pitt exhibits.

Stothert & Pitt and Torrance & Sons stands at the Chemical and Petroleum Engineering Exhibition held at Olympia from 18 to 28 June 1958. Stothert & Pitt exhibited a range of rotary and screw displacement pumps. The associated company of Torrance & Sons Ltd exhibited examples of their mixing, grinding and dispersing units.

Opposite below: Stothert & Pitt stand at Olympia, 1966. A general view of the company's stand at the Ship's Gear International Show, held at Olympia from 6 to 12 July 1966, where items of equipment manufactured by the Crane and Deck Machinery and Pump Departments were shown. Apart from the 'Stevedore 555' electro-hydraulic deck crane, which formed the focal point of the stand, deck machinery equipment included a 7-ton electro-hydraulic mooring winch. The Pump Department was represented by a selection of the pumps of the screw displacement, centrifugal and rotary displacement types destined for the new Cunard passenger liner *QE2*.

Display stand at the British Engineering and Marine Welding Exhibition. Stothert & Pitt regularly exhibited at national and international exhibitions until the 1980s. On display are rotary and positive displacement pumps for dock or deck.

Display stand at German Contractors Exhibition, 1984. In the final years of the company, contractors plant was regularly exhibited at the BAUMA exhibition in Germany where this photograph was probably taken. Two of the last products introduced are shown: the scissors lift and the dumper truck.

nine

People, Playing
and Presentation

As the company expanded, so social clubs and facilities were provided for and by the workforce. In 1903 a rugby team had been established, and by 1940 a football team, bowling and hockey team were also operating. The rugby club players acquired the nickname 'The Cranes' and to this day have as their emblem a crane perched on one leg. An amateur dramatics society was founded and in the 1960s Stothert & Pitt had its own 'beat group', 'The Spectres'.

The *Stothert & Pitt Magazine* reported in great detail the presentations for weddings, retirements or completion of apprenticeship when it began printing in 1948. Normally the various departments would organise the presentation of a set of bathroom scales, a standard lamp or something of the kind, as a gift. The annual Flower and Vegetable Show was always a highlight.

The various departments would also organise annual outings for staff and visits from local educational organisations were encouraged. As well as technical schools or metalworking groups from secondary schools, art groups were also regular visitors. An annual open day for the company allowed friends, family and the general public to see what the workforce got up to.

Of all the social associations, the rugby club 'The Cranes' alone remain.

R.A. Riddles, Chairman, at the Stothert & Pitt Flower and Vegetable Show, 1960. The chairman's duties included the awarding of prizes at the company's annual show.

Above: HRH Princess Margaret visits a Stothert & Pitt stand at the British Industries Fair in Birmingham on 2 May 1951.

Right: Letter to all Stothert & Pitt employees, 1946. After the Second World War a special publication – *Stothert & Pitt 1939-45* – was produced and issued free to every employee along with this letter of appreciation and encouragement.

STOTHERT & PITT, LTD
ENGINEERS.

HEAD OFFICE & WORKS
BATH.

LONDON OFFICE
38 VICTORIA ST. S.W

TITAN FOR BUILDING HARBOURS

MEMBER OF THE
FEDERATION
OF BRITISH
INDUSTRIES

TELEGRAMS: STOTHERT SOWEST LONDON.
PHONE : LONDON ABBEY 1911 (3 LINES).

Telegrams: STOTHERT BATH
PHONE Nº BATH 2277 (4 LINES)

REFERENCES:
OURS:
YOURS:

BATH.
12th October, 1946.

In presenting you with this copy of our War Record I should like to express our appreciation of your services.

In these years of rehabilitation after seven years of war there is a very strong demand for our more peaceful products.

It is incumbent on us to seize this opportunity and accept every contract we possibly can so that we may build up a large circle of important clients in order to see us through the leaner times which may follow. This can only be achieved by the continued loyal co-operation and hard, willing work of us all.

R. B. Pitt

Managing Director.

R.A. Riddles, Chairman of Stothert & Pitt, 1960. Mr Riddles is shown standing alongside the last express steam locomotive designed by him for main-line use in Britain, *Evening Star*. R.A. Riddles had a distinguished career as designer for the London Midland & Scottish Railway and British Railways before becoming chairman of Stothert & Pitt in the late 1950s.

Retirement of Mr S. Freeman from the Pump Shop, 21 July 1960. As was common, a presentation, in this case a mantelpiece clock, was made by foreman Jim Murray. Many photographs of this kind were taken for the *Stothert & Pitt Magazine*.

Dr E. Scott-White. For many years the medical team for the Stothert & Pitt company was headed by the redoubtable Dr Scott-White.

The sickroom at Stothert & Pitt, *c*.1950. Sister E.K. Botting (resident sister, Victoria Works) takes notes, whilst Dr E. Scott-White examines a patient's ear. Nurse E. Cowley (Newark Foundry) stands by to give assistance.

Stothert & Pitt football team, 1958. Although a rugby team bearing the company's name survives to this day, a football team also flourished. In 1958 the team comprised, from left, back row: -?-, J. Anderson, -?-, R. Hudson, K. Leask, -?-. Front row: J. Withers, R. Norris T. Ricketts, -?-, M. Talbot.

STOTHERT & PITT, Limited,
BATH
J. G.

This Pass is the property of STOTHERT & PITT, LTD., and is to be given up if the holder leaves the Company's employment.

It is personal to the employee named and is not to be used by any other person.

R. B. PITT,
Managing Director.

Pass No. 1280

Badge No. 208

STOTHERT & PITT, Limited.

PASS

expiring 31st March, 1945

Dept. 27

Clock No. 110

Name W.V.J. DERRICK

Holder's Signature

Workman's pass issued to Bill Derrick, 1945. Stothert & Pitt was a military target during the Second World War and top-secret projects were undertaken for the Admiralty; however, even in peacetime a workman's identity pass was required.

Visit of careers masters to Stothert & Pitt, 1966. Careers masters from Bath, Frome and Bristol schools toured the works on Tuesday 15 February 1966. The visit was arranged to examine the career prospects at Stothert & Pitt and training schemes available for boys shortly leaving school. An informal discussion and 'question time' followed the tour, and the above photograph shows three of the visitors with Mr T.E.R. Torrance, director and general manager, Crane and Deck Machinery Department (centre right), inspecting a model of the DD2 hydraulic cargo crane.

Party of visiting schoolboys from Monkton Combe School, 5 October 1960. Technical and arts schools, as well as secondary schools, would arrange visits to the works. A range of career opportunities, from apprentice engineer to draughtsman, would be demonstrated to potential employees.

Visiting client Mr Walters, 1969. Mr Walters was a mechanical and electrical engineer with Marequip, and is seen inspecting a train of gear wheels with Mr C.A. Ashman on the left.

Presentation of Awards of Apprenticeship, November 1960. The awards for outstanding apprenticeship took place, as seen here, in the less than salubrious surroundings of the Newark Foundry basement canteen.

Stothert & Pitt rugby team, First XV, 1958/59. The company rugby team had flourished since its inception in 1903. In this view can be seen, from left, back row: K. Russell, W. Mills, A. Shadwell, B. Robbins, A. Allen, R. Cook, D. Hartley, R. Bowler. Front row: G. Shaw, M. Payne, -?-, O. Lonie (Captain), J. Stagg, N. Tucker, J. Perry, -?-.

Presentation to George Webber, senior foreman. With his retirement from the No.2 Machine Shop, Mr Dilwyn Jones, Works Manager, makes an official presentation to George Webber.

The Managing Director of Stothert & Pitt and Mayor of Bath, Councillor Adrian Hopkins, celebrate the manufacture of the 30,000th Vibroll in Abbey Churchyard, Bath, in March 1974.

Opposite above: Visit to Victoria Works by officials from the Port of Bristol Authority, 1951. Alderman Burgess, Mr Matheson and Mr Young of Bristol City Council inspect a turntable assembly for a dockside crane with Col. Pitt of Stothert & Pitt on 25 April 1951. In the same year an order for 3-ton cranes was placed by the Authority. Four cranes survive (preserved) in 2003 at Princes Wharf at the Bristol City Docks.

Opposite below: Fred Harper at Monkswood Reservoir, *c.*1999. Fred Harper, a former employee of Stothert & Pitt, was a retired reservoir-keeper for Wessex Water at the Monkswood Reservoir, St Catherine's, Bath. He is seen operating the light crane, built by Stothert & Pitt in 1898, that was used to move a fine-mesh frame which filtered out large material entering the open-air reservoir. The lifting mechanism of the crane was powered by a 'Pelton' turbine wheel driven by the spring water which filled the reservoir.

Employees of Torrance & Sons, 1928. Ten of the men are: Cecil Morris, front row, sixth from left; Harry Short, front row, extreme right; Sam Allen, second row from front, fourth from right; Harry Howe, back row, sixth from right; Austin Willmott, back row, twelfth from right; Frank Saunders, back row, extreme left; Harry Packer, middle row, extreme right; Percy Nurse, middle row, fifth from right; Leslie Veale, fourth row from front, second from right; and Charlie Harvey, back row, fourth from right.

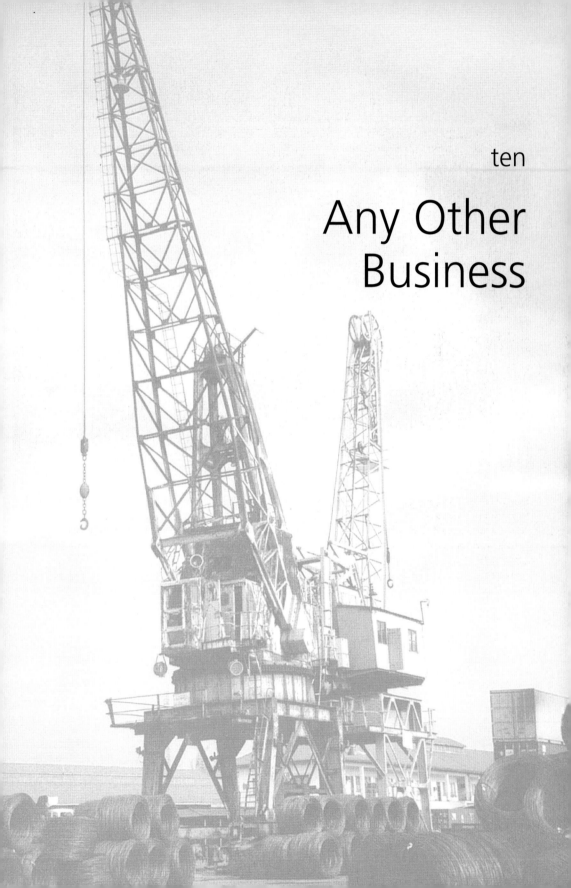

ten

Any Other Business

The photographic department of Stothert & Pitt not only produced photographs for catalogues or serious publicity material, but a great number of illustrations were also created for the staff journal, the *Stothert & Pitt Magazine*, which first appeared in 1948. Photographs for this publication were often produced for their quirkiness or abstract quality rather than their descriptive qualities so often required for publicity material.

The following photographs were all taken for the *Stothert & Pitt Magazine*, which ceased publication in the early 1980s.

Ludlow Waterworks. In 1880 Stothert & Pitt supplied and installed a water-driven turbine, 8ft in diameter, and two variable-stroke double-acting Holman patent pumps, operating at a speed of 23½rpm. This installation was still in operation in 1952 and daily pumped water from the River Teme to a reservoir at Ludlow, 164ft above pump level. The quantity of water being handled was in the region of 190,000 gallons per twenty-four hours.

Deck crane section leaving Newark Foundry, 1963. This massive crane ordered by the Admiralty for installation at their Portsmouth naval base leaves Newark Foundry yard on 19 October 1963.

Rail-mounted crane at Bath Goods Yard, 1925. Stothert & Pitt supplied this crane to the Great Western Railway for their goods yard at Westmoreland Road. Products from the company were trans-shipped here, and this photograph shows components of a crane being handled. A Foden steam lorry attends.

Bridge building at Columbo, Sri Lanka, *c.*1940. This vast crossing is being constructed with the assistance of Stothert & Pitt cranes mounted upon the structure of the bridge itself.

Opposite above: Hand crane on the Scilly Isles, 1963. Many of the company's nineteenth century installations remained in use in more remote locations well into the 1960s. This light-loading crane was still in use when this photograph was taken.

Opposite below: Herbert lathe at Victoria Works Machine Shop, 1948.

Left: Steam engine at the Royal Mineral Water Hospital, 1952. This small engine, used for water pumping at this famous hospital in Bath, bears a Stothert & Pitt makers' plate. It was replaced in 1952 and is now on display at Bath University. The attached plate notes that it was 'reported as manufactured and installed in 1830'. It would appear that the installation of a steam engine at the hot water springs in Bath has been confused with this late nineteenth-century engine which may not have been made by Stothert & Pitt at all but was restored by the company at some point.

Below: Portable scissors lift, 1985. The end of the line! One of the last new products introduced by Stothert & Pitt before closure in 1989 was this scissors lift for use by civil engineering or maintenance contractors.

Above: Gently does it! DD2 cranes at Surrey Commercial Docks, London, are safely lowered into position by the floating crane *London Mammoth* (not made by Stothert & Pitt!) in 1964.

Right: Safe loading indicator for crane, 1940.

Left: Still going strong! 5-ton cranes at Mombasa Harbour, 1999. It was often said that while sailing into any port in the world the first thing you would see would be a Stothert & Pitt crane. Though many have been dismantled or replaced by more modern examples, these cargo-handling cranes in Kenya are still in daily use.

Below: Open Day at Stothert & Pitt, May 1948. The opportunity for the general public to visit the factories and to see the working exhibits was popular with children and adults.

Building Boiler Shops at Victoria Works, 1924. Stothert cranes were, of course, used in any construction work!

The Victoria Works was badly damaged by German bombing over the weekend of 25-26 April 1942. Here an armoured personnel carrier being worked upon is shown amongst the wreckage.

'Touched up' photograph, 1910. The cluttered surroundings of the Victoria Works did not show off photographed products to the best advantage, and so for publicity purposes the background was removed by the photographic department. In this view, a Hercules-type crane, commissioned for Dar-Es-Salaam in Tanganyika, is shown in the process of being 'whited out'.

Mobile concrete mixer, Royal Crescent, Bath, 1980. In the city of Bath the elegant backdrops of Georgian terraces were even used by Stothert & Pitt to advertise their utilitarian products. Most famous of all is the Royal Crescent and here a truck-mounted concrete mixer makes a bright, if slightly incongruous, subject.

Above: A testament to longevity! A 30-ton crane at Salford Docks, 1963. This crane, originally supplied in 1894 as a hand-operated crane, was later converted to steam operation and was still in daily service in 1963.

Next page: Cranes at Bristol Docks. Although some Stothert & Pitt products remain in use in ports in this country, the 1878 Fairbairn-type steam crane seen here in the foreground and four of the cranes in the background have been preserved by Bristol Museum Service as historic monuments. The Fairbairn crane is regularly operated and celebrates its 125th birthday in 2003. The remaining electric travelling cranes on the dockside were being restored in 2003 and one is already in working order.